GREAT CENTRAL IN THE MIDLANDS

GREAT CENTRAL
IN THE MIDLANDS

Images from The Transport Treasury

Compiled by Peter Sikes

GREAT CENTRAL IN THE MIDLANDS

© Images and design: The Transport Treasury 2022. Text: Peter Sikes

ISBN: 978-1-913251-36-9

First published in 2022 by Transport Treasury Publishing Ltd., 16 Highworth Close, High Wycombe HP13 7PJ

The copyright holders hereby give notice that all rights to this work are reserved.
Aside from brief passages for the purpose of review, no part of this work may be reproduced,
copied by electronic or other means, or otherwise stored in any information storage and retrieval system without written
permission from the Publisher. This includes the illustrations herein which shall remain
the copyright of the copyright holder.

www.ttpublishing.co.uk

Printed in Tarxien, Malta by Gutenberg Press Ltd.

Front Cover: BR Standard 9F No. 92113 makes a majestic sight working a Down freight of empty minerals from Woodford to Annesley near Shawell on 24th April 1965. *Ref: MM2825*

Title page: Thompson Class B1 No. 61008 *Kudu* near Whetstone. *Ref: CW10086*

Left: Neasden allocated Standard 5MT No. 73157 works 'The Master Cutler' north through Woodford Halse on 27th May 1957. 'The Master Cutler' was one of only two named trains that ran on the GC main line, the other being 'The South Yorkshireman'. Introduced by the LNER in 1947 and continued by the Eastern Region of British Railways, the train ran between Sheffield Victoria and Marylebone. Departing Sheffield at 7.40am and calling at Nottingham Victoria, Leicester Central and Rugby Central arriving at Marylebone at 11.15am, the return working departing at 6.15pm, arriving in Sheffield at 10.02pm. The downgrading of the GC main line saw the service transferred to Kings Cross in 1958 with the service running for a further ten years before being discontinued in 1968. *Ref: MM258*

Below: Gresley A3 Pacific No. 60111 *Enterprise* at Leicester Great Central shed prepped and ready to take over 'The Master Cutler' from Leicester Central. *Ref: CW11283*

Back cover: Thompson Class L1 2-6-4T No. 67758 passes Leicester South Goods with the 5.30pm Nottingham-Rugby local on 20th June 1960. *Ref: MM1200*

INTRODUCTION

Welcome to 'Great Central in the Midlands'.

I often wonder if the Great Central Railway was ahead of its time by building the line to such a demanding standard or conversely too late in being the last main line constructed to reach London, meaning that it was always playing catch up due to the duplication of routes into the capital. The company had to work hard to win traffic from its rivals and it was only after Samuel Fay joined the company and invested in a policy of heavy marketing and promotion that the railway started to hold its own against its competitors.

Freight traffic was to grow healthily and became the lifeblood of the line, the staples being coal, iron ore, steel, and fish and banana trains. The connection with the Stratford-upon-Avon and Midland Junction Railway at Woodford Halse proved strategically important for freight on the route. Another major centre for freight was at Annesley, Nottinghamshire where in 1947 a service of regular fast freight trains was inaugurated by the LNER to the marshalling yards at Woodford Halse. The famous 'runners' as they were known by the enginemen ('windcutters' by enthusiasts) carried coal from the Nottinghamshire coalfield and steel products from Yorkshire and the North East destined for the London area or South Wales.

Despite the previous paragraphs this book is not intended to be a written history of the route but a pictorial journey predominantly spanning the British Railways era under Eastern Region control until the run down of the line after transfer to the London Midland Region in January 1958, through to its complete closure in September 1969.

This photographic selection, predominantly from the camera of Mike Mitchell, follows the line from Staveley Central through the East Midland cities of Nottingham and Leicester to Brackley Central in Northamptonshire, the southernmost station on the Great Central main line in the Midlands area.

Large Director Class D11/1 No. 62661 *Gerard Powys Dewhurst* pictured on the approach to Rothley station on the 7.45pm Nottingham-Leicester on 26th June 1959. Similar scenes can still be witnessed today. *Ref: MM888*

Living near to Leicester I often drive into the city along the A6 where there are constant reminders that the railway once existed. The line reached Leicester Central station on a high embankment and there are several blue brick bridge buttresses extant, and where the line crossed the River Soar there is a long stretch of brick arches that are today home to several small businesses, unfortunately there is no prospect of trains rumbling above them ever again. It also means I am fortunate that the 8½ mile preserved section of the GCR is on my doorstep and it is evident that the quality of each construction on the line was built to the highest of standards. It meant that goods and parcels trains, as well passenger trains, were usually able to run at high speeds, this being attributable to the quality of the line formation and alignments requiring few speed restrictions. It is often mistakenly quoted that the line was built to a 'continental loading gauge', a myth that seems to have been perpetuated over the years despite the fact it didn't exist when the line was constructed! It was, in fact, built to the standard MSLR (Manchester, Sheffield and Lincolnshire Railway) loading gauge.

As with the preserved GCR of today there are a large variety of classes of locomotive featured, especially at Nottingham where the wonderful cathedral-like Victoria station initially shared services with the Great Northern Railway, until they came together with the grouping of 1923 when the route came under the auspices of the London and North Eastern Railway. Then there were the cross-country services that joined the GCR main line at Culworth Junction bringing Great Western engines as far north as Leicester and Nottingham plus the introduction of Midland type classes transferred as they neared the end of their working lives. The sad demise of the line is evident in many of the photos as we enter the second half of the 1960s, the smaller stations closing and just being left to rot as traffic declined and the increase in road building gave the public another way to get to their destination.

Peter Sikes, Syston, Leicestershire – August 2022.

Thompson Class B1 4-6-0 No. 61008 *Kudu* at Nottingham Victoria at the head of 'The South Yorkshireman' bound for Marylebone on 8th March 1958. A version of this service runs today on the preserved Great Central Railway. *Ref: MM491*

Staveley Central

Opened on 1st June 1892 by the Manchester, Sheffield & Lincolnshire Railway (MS&LR) as Staveley Town but renamed Staveley Central on 25th September 1950 by British Railways, to reduce confusion with the ex-MR station, also called Staveley Town, which was about 250 yards away on the same street. The renaming also reduced the likelihood of people confusing the station with that at Barrow Hill and Staveley Works, which even though was commonly referred to by its former, was officially renamed Barrow Hill on 18th June 1951 to further differentiate the stations.

The station was the northern junction for the loop line to Chesterfield Central comprising four platforms. The timber-built booking hall was on the Lowgates Road overbridge and facilities included a waiting room on each platform. The station was also the junction for branches to the Ireland, Hartington and Markham collieries and at the south end was Staveley (GC) Engine Shed (code 38D, changing to 41H in BR days). This, too, was subject to confusion with the ex-MR Staveley engine shed over a mile away at Barrow Hill. Staveley Central closed on 4th March 1963, but continued to serve summer weekend excursion traffic until the end of the 1964 season.

Above: Gresley A3 No. 60111 *Enterprise*, pictured at Staveley Town on an unknown date in 1957. *Ref: NS205660*

Top Right: Robinson Class J11 0-6-0 No. 64433 passes through Staveley Central 1957. *Ref: NS205660A*

Bottom Right: Class O1 2-8-0 No. 63886 passes through Staveley Central in April 1962. The O1s were introduced in 1944 and were a Thompson rebuild of Robinson's GC design with a 100A boiler and Walschaerts valve gear. *Ref: AF0648*

GREAT CENTRAL IN THE MIDLANDS

GREAT CENTRAL IN THE MIDLANDS

Chesterfield Central

Located on the GCR Chesterfield Loop which ran between Staveley Central and Heath Junction (just north of Heath railway station). The station opened in 1892 and was closed in 1963. Although the official closing date was 4th March 1963, the last passenger train to use the station did so on 15th June, when *Flying Scotsman* stopped there during a Railway Preservation Society tour from Sheffield Victoria to Marylebone. It remained open for goods traffic until 11th September with a private siding continuing in use after that.

Compared to nearby Chesterfield Midland, the station was little used. The number of passengers using the station during the week ending 19th August 1961 was 1,829, in the same week Midland station was used by 22,285 passengers. The station was demolished by 1973 to make way for Chesterfield's inner-relief road, much of which was built along the former trackbed.

Left: A pair of D11/1 Robinson 4-4-0 'Large Director' class locos dating from 1920 are pictured at Chesterfield Central. On the left is No. 62666 *Zeebrugge* with No. 62667 *Somme* on the right, both appear ready to depart with stopping passenger trains. Chesterfield's famous twisted spire is easily identified on the skyline. *Ref: NS202615*

Above: Thompson B1 4-6-0 No. 61312 arrives at Chesterfield Central with an express working. *Ref: NS205664*

Above: Thompson B1 4-6-0 No. 61047 arrives at Chesterfield Central with the 1.26pm stopping service for Nottingham on 17th December 1960. Built by the North British Locomotive Company in Glasgow it entered service for the LNER as No. 1047 on 3rd June 1946 and was allocated to Norwich Thorpe. Although the loco looks in need of attention in the photo above, it would not be withdrawn until 16th September 1962 from Sheffield Darnell, eventually being cut up at Cashmore's the following year.
Ref: H1964

Right: Class O1 2-8-0 No. 63868 passes through Tibshelf station with a long rake of mineral wagons sometime in 1965. Originally designed and built by J. Robinson as Class O4 for the War Department, it entered service in November 1919. The O1s were introduced in 1944 by Thompson being rebuilt with a 100A boiler, Walschaerts valve gear and new cylinders. 63868 was converted and entered service in February 1945. The locomotive would be withdrawn in July 1965 and cut up by Drapers of Hull in November of the same year. *Ref: NS205655*

Tibshelf

Opened on 2nd January 1893 by the Manchester, Sheffield and Lincolnshire Railway, the station was originally named Tibshelf Town, reflecting the village's optimistic aspirations to obtain official town status. Tibshelf never did become a town but the station kept this name throughout its operating life. The line ran through a mainly industrial landscape dominated by mining. To the north of the station was a deep cutting where a tunnel was originally intended; fears of damage through mining subsidence forced the change in the plans. Tibshelf High Street crossed a bridge over this cutting.

To the south the line crossed over the Midland Railway's branch line from Westhouses & Blackwell (on the Erewash Valley Line), to Mansfield Woodhouse, just to the east of their own Tibshelf & Newton station. This line closed to passengers on 28th July 1930 but remained in use for freight and coal trains long afterwards; the route is now a footpath but Tibshelf & Newton station still stands. Tibshelf Town station closed on 4th March 1963. The collieries also closed around this time, although coal mining continued to be a major source of employment for the village, with around 2,000 of Tibshelf's inhabitants still working at local pits as late as the 1980s.

The route of the Great Central line was redeveloped by Derbyshire County Council in the late 1960s and early 1970s, and now forms part of the Five Pits Trail network.

GREAT CENTRAL IN THE MIDLANDS
Kirkby Bentinck

Kirkby Bentinck, near Kirkby-in-Ashfield, Nottinghamshire was on the Annesley branch of the Manchester, Sheffield and Lincolnshire Railway on the section from Nottingham Victoria to Sheffield Victoria. The station was opened in March 1893 and closed in March 1963 after 70 years in service. Until 1st August 1925 it was named Kirkby & Pinxton station.

Above: Thompson Class L1 2-6-4T No. 67799 works the 10.9am Sheffield-Nottingham, near Kirkby Bentinck on 18th July 1959. *Ref: MM927*

Left: Thompson Class O1 2-8-0 No. 63901 with just a brake van in tow pictured at Kirkby Bentinck on 18th July 1959. *Ref: MM928*

Right: On 8th October 1966, over three years after closure, Kirkby Bentinck station looks in good shape apart from the weeds proliferating the platforms. *Ref: MM3294*

Hucknall Central

The station opened as Hucknall Town on 15th March 1899. Constructed by civil engineering contractors Logan and Hemmingway, the firm had been awarded a number of contracts by the GCR's predecessor the Manchester, Sheffield and Lincolnshire Railway, including expansion of the company's No.1 Fish Dock at the Port of Grimsby in the 1870s, parts of the company's London extension and the new station in Nottingham. Soon after grouping, as part of the London and North Eastern Railway, the station was renamed Hucknall Central on 1st June 1923. The station was closed to both passengers and goods on 4th March 1963.

Above: Hucknall Central viewed from a train on 1st October 1960. *Ref: 000H1882*

Right: Bulwell Common opened with the line on 15th March 1899. The choice of 'Common' in the name has remained something of a puzzle, as Nottingham City Council have no record of any common land ever having been designated in the Bulwell area.

The station was the fourth to serve the northern Nottingham suburb of Bulwell, directly or indirectly, following the Midland Railway station later known as Bulwell Market on their line from Nottingham to Mansfield and Worksop, the Great Northern Railway (later LNER) station called Bulwell Forest on their own route up the Leen Valley and the same company's Basford and Bulwell station (later renamed Basford North), on their Derbyshire and Staffordshire Extension to Ilkeston, Derby, Uttoxeter and Stafford. To the south of the station, a west-to-north connecting curve from the GNR's Basford and Bulwell station joined the GCR at Bulwell South Junction, while to the north, a connecting spur to the GNR's Leen Valley line branched off north-eastwards at Bulwell North Junction; further north still was the lengthy Bulwell Viaduct which crossed over the Midland line and spanned the Leen Valley, while a short distance to the north of this was a fifth station to bear the name Bulwell – the relatively short-lived Bulwell Hall Halt (24th April 1909-5th May 1930).

Bulwell Common station closed to passengers and goods on 4th March 1963. Although the Station Master's house still remains the station site was redeveloped for housing.

Above right: Class B16/3 No.61448 passes through Bulwell Common on a mixed freight on an unknown date. This class of locomotive was a Thompson rebuild of the Raven Class B16/1. No. 61448 was built in June 1923, its first allocation being Gateshead, and had a 41 year working life, being withdrawn in June 1964 and cut up three months later at Hughes Bolkcows of North Blyth. *Ref: NS205651*

Below right: G. E. Hill Class N7/3 0-6-2T No. 69692 calls at Bulwell Common with a local service on an unknown date. As can be seen in this photograph there were extensive sidings conveniently situated between Bulwell North and South junctions, the sidings handled freight to and from the Great Northern's Derby and Staffordshire lines. *Ref: NS205652*

GREAT CENTRAL IN THE MIDLANDS

Bulwell Common

GREAT CENTRAL IN THE MIDLANDS

Nottingham Victoria

Above: Designed by H. A. Ivatt in 1909, Class J5 0-6-0 No. 65494 at Nottingham Victoria in August 1952. (An unidentified Gresley J6 stands in the adjacent platform. *Ref: AF0322)*

Nottingham Victoria railway station was a joint venture between the Great Central Railway and Great Northern Railway. It was designed by the architect Albert Edward Lambert, who also designed the rebuild of the Nottingham Midland station, and was opened by the Nottingham Joint Station Committee on 24th May 1900, closed on 4th September 1967 by the London Midland Region of British Railways. Apart from the clock tower the station building was entirely demolished being replaced by the building of the Victoria Shopping Centre.

The station's construction was on a grand scale: a 13-acre site was acquired at a cost of £473,000 (£56.1 million) in the heart of Nottingham's city centre, negotiations for which had taken three years to acquire the land. The construction called for the demolition of some 1,300 houses, 24 public houses, and St. Stephen's Church, Bunker's Hill, following which about 600,000 cubic yards of sandstone was excavated from the site. Measuring about 650 yards in length from north to south the site had an average width of 110 yards with a tunnel at each end for access.

The Great Central Railway and Great Northern Railway shared the station, splitting into two lines at Weekday Cross Junction. The two owners failed to reach agreement on the station's name; the Great Central naturally wanted it called 'Central', a proposition the Great Northern, still smarting from the incursion into its territory made by the London Extension, would not accept. The two railway companies operated separate booking offices, the Great Central issuing tickets reading 'Nottingham Central', whilst the Great Northern's window bore the name 'Nottingham Joint St'n'. In the 1898 Royal Atlas of England, the station was recorded as the 'Grt Central & Grt Northern Joint Central Station'. The town clerk resolved the situation by suggesting the name 'Nottingham Victoria' to reflect the fact that the planned opening date coincided with Queen Victoria's birthday; a suggestion that was readily accepted at a meeting of the Nottingham Joint Station Committee.

Nottingham Victoria station was officially opened without ceremony on 24th May 1900, over a year after the commencement of services on the new railway line. The main station building was in true Victorian splendour. It was constructed in a Renaissance style using the best-quality red-faced bricks and Darley Dale stone, with space at the front for Hackney carriages, which was covered by a canopy.

An evocative image of H. A. Ivatt Class J6 0-6-0 No. 64200 leaving Nottingham Victoria with a southbound service. The bridge adjacent to the signal cabin carries Lower Parliament Street over the railway. *Ref: CW11400*

Above: Introduced into service with the LNER during May 1931, Gresley Class K3 2-6-0 No. 61914 waits on the middle road at Nottingham Victoria. Although no date was given with the photograph the shed code shows 38A Colwick, which is where the locomotive was allocated between June 1954 and August 1960 before being moved to Immingham (40B). No. 61914 would return to Colwick in September 1961, now changed to 40E, from where it was withdrawn on 23rd August 1962. *Ref: CW11546*

The station comprised two large island platforms, as can be seen in the photograph above, approximately 1,250 feet long, with two bays at each end for local traffic, giving a total of 12 platform faces. The island platforms were well equipped, each having dining and tea rooms together with kitchens, sleeping facilities for staff, waiting rooms and lavatories. All of these buildings were lined with buff-coloured glazed tiles embellished with a chocolate dado. Pillars measuring 42 feet 6 inches held up an enormous three-part glazed canopy measuring 450 feet in length, with a centre span of 83 feet 3 inches and a pair of flanking spans each of 63 feet 9 inches. Additional glass roofs were also provided over the double-bay platforms. There was an electrically lit subway system covering the breadth of the station, frequently used for transporting luggage, thereby avoiding the need to carry it over the footbridges. The subway was linked to the main station by four lifts serving the booking hall, cloakrooms, and the two island platforms. The station had freight passing loops, north and south signal cabins, and two turntables.

The traffic that passed through was varied, including London–Manchester expresses, local and cross-country services – such as York to Bristol via Oxford – as well as freight workings. As the station was shared with the Great Northern Railway (already well established when Victoria opened), a large network of lines serving many destinations was available from Victoria. The new services offered unbeatable competition to the Midland Railway. Although the Midland offered more trains per day to Leicester, the journey time of 40 minutes was outclassed by the Great Central's time of 27 minutes non-stop. The journey time north to Sheffield was similarly impressive at 50 minutes.

A wonderful view of Thompson B1 4-6-0 No. 61008 *Kudu* at the south end of Nottingham Victoria, possibly taken from Lower Parliament Street bridge, at the head of a southbound express for Marylebone. 61008 carries a 38C shed plate which was Leicester (G.C.) until 1958, which would date the picture as being taken somewhere between 1956 and 1958. The scene is full of interest with plenty of loco spotters on the platform and a wheeltapper checking the carriage wheels before departure, having walked past Thompson Class L1 2-6-4T No. 67752 waiting on the middle road with an unidentified loco. The south turntable can be seen above the carriages on the right tightly hemmed in by the retaining wall with Glasshouse Street above.
Ref: NS205646

A view from the north end of Nottingham Victoria of three local workings under preparation on 29th June 1957. From left to right we see Peppercorn Class K1 2-6-0 No. 62046 of York (North) shed, H. A. Ivatt Class J6 0-6-0 No. 64225 and Robinson Class A5 4-6-2T No. 69810, the latter two engines both had long working lives of 44 and 46 years respectively.
Ref: MM311

GREAT CENTRAL IN THE MIDLANDS

Riddles Class 9F 2-10-0 No. 92092 has just exited Mansfield Road Tunnel and passes through Nottingham Victoria on a mixed freight sometime in 1959. *Ref: NS205650A*

Another shot of a 9F passing Nottingham Victoria North signal cabin, this one of No. 92043 with its train of steel girders on 1st October 1960. Both photographs show to good effect the sandstone rock faces that formed the cavernous bowl that was excavated to build the station. *Ref: H1884*

The R.C.T.S. Great Central Rail Tour (1X21) ran on 13th August 1966. The train comprised eleven coaches and was run to commemorate the loss of through passenger workings over much of the Great Central London extension. The service cuts were to take effect on 5th September 1966. As seen in the picture above No. 34002 *Salisbury* came off the train at Nottingham Victoria after departing London Waterloo and joining the GC at Aylesbury. The loco spent the afternoon at Colwick MPD.

The train was taken forward by Stanier 8F No. 48197 and this portion toured north Nottinghamshire ending up at Shireoaks. From here the train was taken over by Thompson B1 No. 61131 and travelled via Rotherham Central to Elsecar Junction. Electric traction in the shape of BR Class EM1 No. E26053 *Perseus* was then used to take the train forward to Sheffield Victoria via Penistone. No. 61131 then rejoined the train which travelled directly on the GC main line to return to Nottingham Victoria, where 34002 *Salisbury* was waiting to return the tour to the capital, terminating at Marylebone.

The arrival at Marylebone was over one hundred minutes late. The majority of this delay was attributed to lack of adequate time given to the need for a double reversal at Neasden Junction early on in the first leg of the tour. The timing sheet allowed 3 minutes for what reportedly took about 45 minutes. Unable to make the time up the tour ran late for the rest of the day.

Above: Bulleid West Country Class 6MT 4-6-2 No. 34002 *Salisbury* moves off the stock of the RCTS Great Central Rail Tour at Nottingham Victoria on 13th August 1966. *Ref: LRF8502A*

Right: A commendably clean Stanier 8F No. 48197 waits to take over the tour and will run through to Shireoaks (a former pit village) in Nottinghamshire. Note that the shed plate has been replaced with the painted letters COLK (for Colwick), which also appears on the buffer beam. The popularity of this tour is evident by the crowded platforms. *Ref: LRF8500*

Another Great Central Rail Tour ran on 3rd September 1966 and followed a similar itinerary to that described on the previous pages. This time the train was brought up from the capital to Nottingham Victoria by Merchant Navy No. 35030 *Elder Dempster Lines*. From here the train was taken over by Thompson B1s Nos. 61173 and 61131 this time travelling to Elsecar Junction via Woodhouse East Junction and Rotherham Central. Electric traction took over again, this time by BR Class EM1 No. E26021 to Sheffield Victoria. The two B1s then returned to Nottingham Victoria but via Killamarsh Junction, Mansfield Central, Annesley North Junction and Bagthorpe Junction. The tour arrived at Marylebone at 21.37 against a booked time of 21.20.

Above: Bulleid Merchant Navy No. 35030 *Elder Dempster Lines* waits to return the LCGB Great Central Rail Tour to Marylebone from Nottingham Victoria on 3rd September 1966. Sulzer Type 2 (later class 24) No. D5000 can be seen in the adjacent platform. *Ref: LRF8516*

Right: Thompson B1s Nos. 61173 and 61131 are pictured by the south turntable after working the LCGB Great Central Rail Tour back to Nottingham Victoria from Sheffield Victoria on 3rd September 1966. *Ref: LRF8514*

GREAT CENTRAL IN THE MIDLANDS

Above: Riddles Standard Class 9F 2-10-0 No. 92031 works an Up fitted freight and is pictured leaving the southern outskirts of Nottingham on 18th February 1958. Through the mist on the left can be seen the chimneys of Wilford Power Station. *Ref: MM469*

Left: Our final picture from Nottingham Victoria shows examples from two of arguably the most famous British locomotive designers. Stanier Jubilee Class 6P 4-6-0 No. 45608 *Gibraltar* of 55A Leeds (Holbeck) awaits departure alongside Gresley Class K3 2-6-0 No. 61943 at the north end of the station. Although withdrawn in September 1962, No. 61943 was transferred to stationary boiler duties at Colwick shed (40E) and didn't succumb to the cutter's torch until 1966. *Ref: NS205645*

Gotham Sidings

Gotham Sidings were located where the line crosses Gotham Moor in Nottinghamshire, between Rushcliffe Halt and Ruddington railway station. The sidings were constructed between 1899 and 1900, after the completion of the London Extension in 1899, to serve a short branch line to the Limestone (plaster) works of the Gotham Company's works (later British Gypsum) situated in Gotham itself. The branch line closed in 1969.

Above: Henry Ivatt Class J6 0-6-0 No. 64257 pictured while shunting wagons at Gotham Sidings on 18th May 1957. *Ref: MM250A*

Below: Another Ivatt J6, No. 64248, makes its way towards Gotham with five wooden bodied mineral wagons on 4th March 1958. This class of locomotive had long working lives with the examples shown here both surviving for over 41 years. *Ref: MM486*

Above: On 10th September 1960 the East Midlands Branch of the R.C.T.S. ran 'The Gypsum Mines Tour'. Starting from Kegworth station sidings the train, hauled by Peckett Class R2 0-4-0ST Lady Angela, carried passengers in four wooden bodied open wagons with a box van bringing up the rear, to the Gypsum Mines Ltd. works at Kingston-on-Soar. *Ref: MM1304*

Below: Passengers were forwarded to Gotham Public Siding by buses where they would find Ivatt 2MT 2-6-2T No. 41280 of Annesley shed (16D) waiting for them. The train is pictured waiting to cross the road at Gotham Works. On this part of the journey the passengers were given a bit more room, the train comprising an ex-LMS brake van, five open wagons and three BR 20 ton brake vans. They would detrain at Gotham Junction into three ex-LMS coaches for the onward journey to Nottingham Queens Walk Goods. *Ref: MM1307*

GREAT CENTRAL IN THE MIDLANDS
Rushcliffe Halt

Rushcliffe Halt was a late arrival on the Great Central Railway, opening on 3rd July 1911 to serve Rushcliffe Golf Club. Due to a lack of room the station was built in a cutting with two platforms instead of the preferred island platform layout. The station had no goods service but a passing loop and sidings were later provided for the East Leake gypsum mine which had opened in 1903 and is still operating today. A signal cabin was added to the north of the station in the 1940s to control the sidings.

The Halt closed to passengers in 1963 although freight continued to serve the British Gypsum works until the early 1980s. A chord was constructed at Loughborough from the Midland Main Line to maintain access with freight traffic returning to the gypsum works in 2000. At the time of writing freight traffic has been suspended due to a defective bridge that crosses the A60 near the Brush Works at Loughborough.

Above Left: Annesley-based Riddles Standard Class 9F 2-10-0 No. 92092 works the 6.5pm Nottingham-Rugby stopping service past the plasterboard works at Hotchley Hill, near Rushcliffe Halt, on 19th May 1964. The aforementioned signal cabin can be seen on the left as a northbound local service passes. *Ref: MM2461*

Below Left: A view of Rushcliffe Halt looking south on 19th May 1964. *Ref: MM2463*

Above: Stanier Black Five 4-6-0 No. 44825 speeds through the northbound platform at Rushcliffe Halt working the 4.38pm Marylebone-Nottingham on 18th June 1966. As can be seen on the smokebox door the locomotive was at this time allocated to Colwick. Built at Derby Works in 1944, its first allocation was to Nottingham (16A) and it was withdrawn from Carlisle Kingmoor (12A) in October 1967 after a working life of just under 23 years. *Ref: MM3157*

GREAT CENTRAL IN THE MIDLANDS

East Leake

East Leake station is the only surviving Great Central Railway station that was accessed from an underbridge rather than an overbridge. The station opened on 15th March 1899 and closed on 5th May 1969. Situated in an attractive wooded area on the outskirts of the village the station had a small goods yard to the north, this was redeveloped for housing in the 1990s. The station buildings have long been demolished, with the rubble used to fill in the area from the road entrance to the platform. The island platform has remained in situ.

Left: Robinson Class O1 2-8-0 No. 63689 is pictured on the approach to East Leake with a fitted freight on the evening of 14th June 1957. The loco carried several numbers during its working life of 44 years and 8 months. Entering service in March 1918 as ROD Class O4/3 No. 1671, she was then loaned to the LNWR in 1920 and numbered 2971, taken into LNER stock on 31st May 1924 as No. 6341 then rebuilt as a Class O1 in 1945, renumbered 3689 in November 1946 with the final number being applied by British Railways but not until October 1949. *Ref: MM295*

Above: Robinson Class A5 4-6-2T No. 69812 is pictured near East Leake on 12th February 1958, with what appears to be an inspection saloon, not too taxing for a locomotive classified 4P. A Colwick based engine first introduced by the GCR in 1912 and allocated to Neasden, she would be withdrawn from service in July 1959. *Ref: MM458*

Above: Robinson Class O1 2-8-0 No. 63886 pictured south of East Leake working an Up freight on 12th February 1958. *Ref: MM459*

Right: Gresley K3 2-6-0 No. 61982 departs East Leake station working the 4.15pm Nottingham Victoria-Leicester Central on 13th May 1961. The train is entering the spacious cutting at the south end of the station, which during its construction saw approximately 330,000 cubic yards of spoil removed, this was then used to create the embankment near to the River Soar at Loughborough. *Ref: MM1440*

GREAT CENTRAL IN THE MIDLANDS

GREAT CENTRAL IN THE MIDLANDS

Barnstone

Approximately a mile south from East Leake station the line passed through a short brick arch tunnel measuring 110 yards, it was crossed by Rempstone Road. Known simply as Bridge 314 – Barnstone Tunnel.

Above Left: Gresley Class V2 2-6-2 No. 60911 working 'The South Yorkshireman', pictured from the north portal at Barnstone Summit on 12th September 1959. *Ref: MM1034*

Below Left: An unidentified Stanier Black Five works the 5.15pm Nottingham-Marylebone and is seen on the approach to the north portal of Barnstone Tunnel on 4th May 1963. *Ref: MM2105*

Above: Riddles Class 9F 2-10-0 No. 92092 works a Down freight through Barnstone Tunnel on 4th May 1963. *Ref: MM2106*

Above: Stanier 5MT 4-6-0 No. 45301 works the 10.20am (SO) Hastings-Sheffield at Barnstone Summit on 24th August 1963. *Ref: MM2287*

Left: Riddles Standard Class 5 No. 73071 on the 12.30pm (SO) Rugby-Nottingham stopping service on 11th July 1964. The locomotive was at this time allocated to Woodford Halse (2F). *Ref: MM2552*

Riddles Standard 9F 2-10-0 No. 92090 works an Up Annesley-Woodford 'runner' pictured leaving the south portal of the short tunnel at Barnstone Summit on 22nd August 1964.
Ref: MM2683

GREAT CENTRAL IN THE MIDLANDS
Stanford-on-Soar

Stanford Viaduct straddles the Nottinghamshire/Leicestershire border, a viaduct was required here to cross the River Soar. Constructed between 1896 and 1898, the GCR contracted the build to Henry Lovatt of Wolverhampton who built the graceful multi-arched viaduct out of blue brick. The three central arches are skewed to allow the Soar to pass underneath. The contractor was responsible for the section of the railway from East Leake to Aylestone.

Above Left: Standard 4MT 2-6-0 No. 76043 heads south over Stanford viaduct on the 6.15pm Nottingham-Marylebone service on 27th August 1961. *Ref: MM1644*

Above: The viaduct is seen to good effect in this view of Thompson B1 4-6-0 No. 61028 *Umseke* crossing and heading towards Loughborough Central with the 10.25am Poole-Bradford (SO) on 9th September 1961. *Ref: MM1649*

Below Left: Stanier 8F 2-8-0 No. 48211 pictured on an engineer's working near Stanford-on-Soar on 8th June 1963. *Ref: MM2165*

Loughborough Central

Loughborough Central was opened on 15th March 1899, and was built to the standard GCR arrangement of having an island platform set between the two main running lines. The platforms are 400 feet long, capable of accommodating trains of up to six coaches. The station buildings are unique on the preserved railway, the only station with a complete canopy, which is also the longest in railway preservation.

As most of us are aware the station was reopened as part of the restored heritage railway in 1974. Today it sees much activity with services running south from the station to Leicester North (formerly Belgrave & Birstall). The station buildings, original GCR signal cabin sited to the north, and the three original water tanks are all Grade II listed.

Above: Thompson B1 4-6-0 No. 1248, pictured shortly after passing through Loughborough Central, crosses over the Midland Main Line at Loughborough Midland station, working a Down unfitted freight in December 1948. Although approaching the first anniversary of nationalisation, the locomotive still carries its LNER livery. Of note is the large advertising sign extolling the courses available at Loughborough College. *Ref: NS205637*

Top right: With approximately 60 mineral wagons in tow, Riddles Class 9F 2-10-0 No. 92093 is pictured on a southbound 'runner' near Loughborough on 9th September 1961. *Ref: MM1651*

Below right: Gresley Class V2 2-6-2 No. 60890 on an Up express freight passing through Loughborough Central on 12th March 1958. The sidings on the left no longer exist, being given over to redevelopment. The station is now a hive of activity, the track to the right of the loco is used to store mainly diesel locomotives and carriages. The area to the right of the signal cabin, which can be seen between the train and the building (now used as the operations centre), is where the engine shed is located, multiple locos can be seen here at any one time with several either being maintained, overhauled or undergoing restoration. *Ref: MM497*

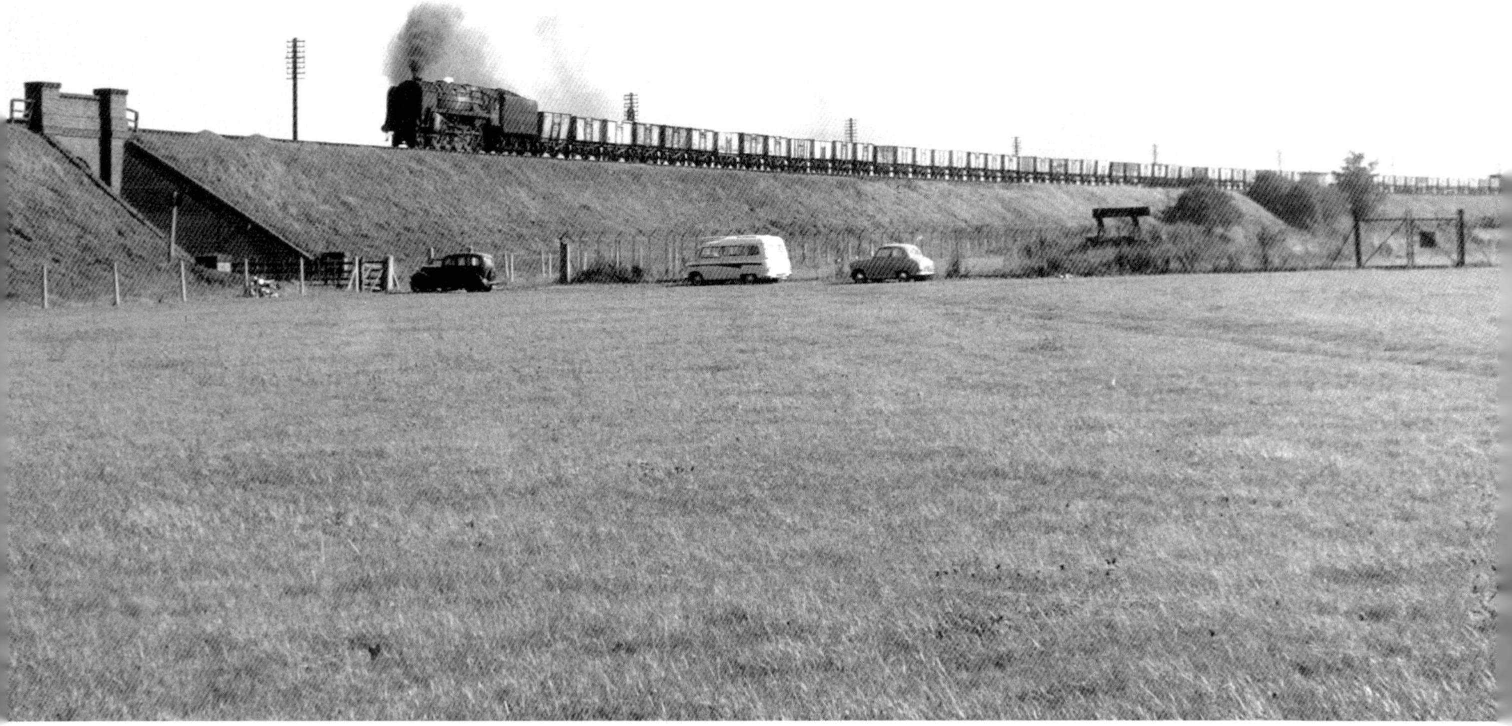

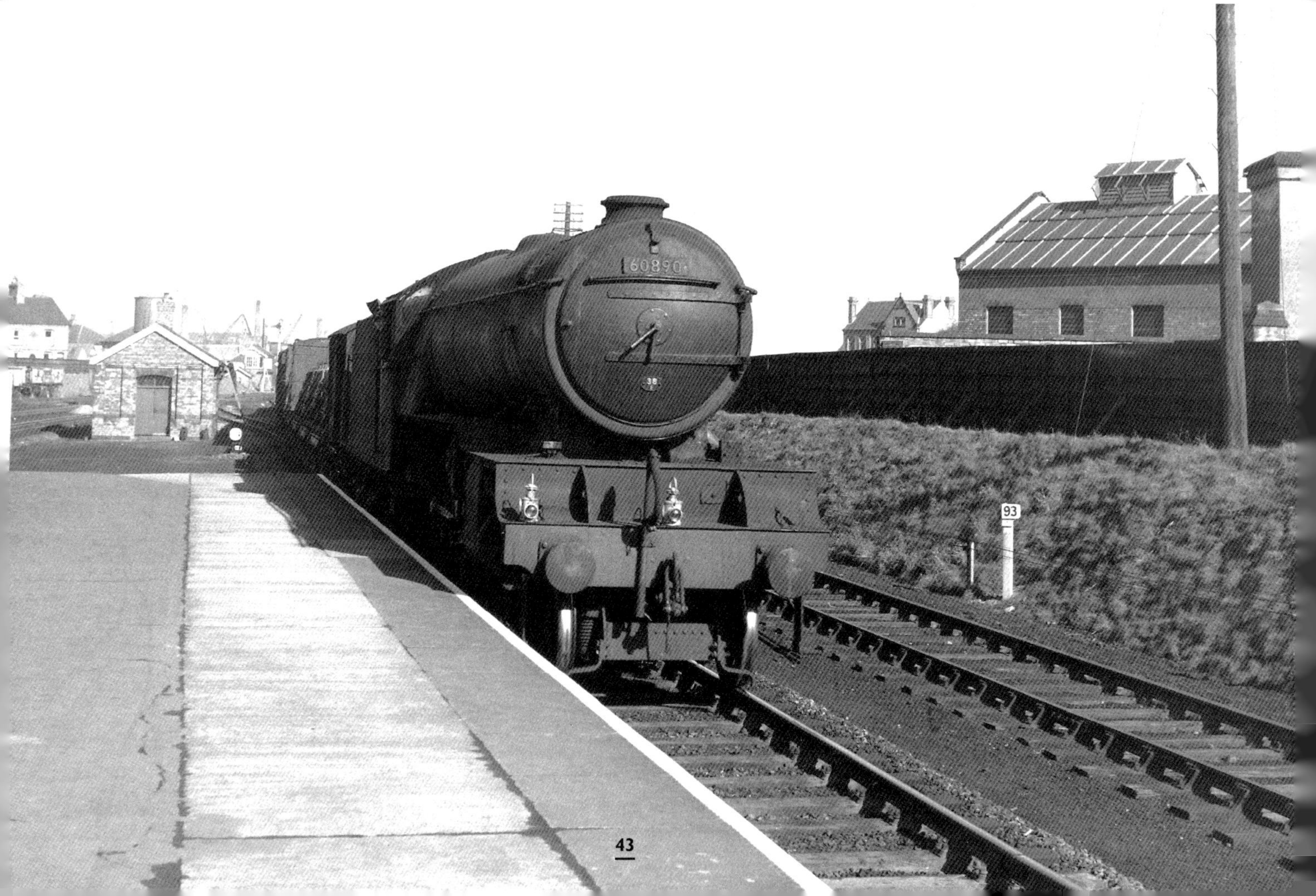
60890
93

Crosti-boilered 2-10-0 No. 92021 works an Up through freight near Loughborough on 18th April 1964. One of the ten Standard 9Fs that were experimentally built with Franco-Crosti boilers. In the event, the experiment did not deliver the hoped-for benefits, and efficiency was not increased sufficiently to justify the cost, also conditions were unpleasant on the footplate in a cross-wind, this despite the later provision of a small deflector plate forward of the chimney. These problems led to the subsequent sealing off of the pre-heater drum between 1959 and 1961, the locomotives were then worked conventionally. As a result of this, there was a reduced ability to generate steam, and so their power classification was reduced from 9F to 8F. *Ref: MM2414*

Quorn & Woodhouse

Thompson B1 4-6-0 No. 61085 approaches Quorn & Woodhouse station with the 9.25am Nottingham to Leicester service on 24th July 1961. Track maintenance continues as the railway is running its normal service and freshly laid ballast can be seen stretching back towards Loughborough. *Ref: MM1571*

Quorn & Woodhouse railway station served the three villages of Quorn, Woodhouse and Woodhouse Eaves in Leicestershire. Well-known to many of us travelling on the preserved section of the GCR, it is the first station reached after departure south from Loughborough and possesses a large station yard which comes into its own at the railway's gala events.

Quorn is laid out to appear as it would have been in the 1940s, that of a typical rural LNER station. The station is grade II listed and has a number of attractions, including the 1940s era NAAFI Tea Room situated underneath the station road bridge, a period Station Master's office, as well as wartime films showing in one of the waiting rooms. The signal cabin to the south of the station is from Market Rasen and is a Manchester, Sheffield & Lincolnshire design from circa 1886 – making it the oldest structure on the railway. Adjacent to the signal cabin a turntable has been installed, although the original goods yard did not possess one. It was built in 1909 by Cowans Sheldon Ltd of Carlisle, and was first installed at York.

During the Second World War the War Ministry expanded the goods yard and added extra sidings. It was considered that the countryside around here was reasonably safe from the attention of enemy bombers. Ammunition was stored in bases in the local area during the build up to D-Day. Beaumanor Hall at nearby Woodhouse was an outpost of Bletchley Park Station X, important in code-breaking work, such as the Enigma codes.

GREAT CENTRAL IN THE MIDLANDS

Above: Fairburn 2-6-4T No. 42252 pictured near Quorn & Woodhouse on the 6.50am Woodford-Nottingham local on 13th June 1964. *Ref: MM2495*

Left: A scene not too dissimilar to the one that will greet you today, and that includes the locomotive. Riddles Standard 5MT 4-6-0 No. 73156 pulls into Quorn & Woodhouse station at the head of the 9.30am Leicester-Nottingham on 24th July 1961. *Ref: MM1566*

Built in 1956, No. 73156 is the only surviving BR Standard locomotive that was constructed Doncaster Works. For much of her brief life in BR service, '156' was allocated to ex-Great Central depots, initially Neasden in December 1956. She graduated to the North West later on via several sheds in the West Midlands by then under LMR control, finally being withdrawn from service whilst at Bolton MPD in November 1967. Accumulated mileage during eleven years of service was approximately 325,000. Owned by Bolton Steam Locomotive Co. Ltd, the loco is now a popular and regular performer on the preserved GCR.

GREAT CENTRAL IN THE MIDLANDS

Rothley

Above Left: Rothley station as it was on 2nd July 1960. The station looks similar today, restored as near as possible in a pre-First World War Edwardian style with gas lighting on the platforms. The major change to that seen in the photograph is the location of the signal cabin (acquired from Blind Lane on the Neasden to Northolt line) which is now positioned opposite the Gentlemen's toilet. *Ref: H1598*

Below Left: An unidentified Fowler 4P 2-6-4T pictured at Rothley station with the 5.30pm Nottingham-Rugby in the platform on 9th July 1960. *Ref: MM1228*

Above: Thompson Class B1 4-6-0 No. 61142 draws into Rothley with the 5.5pm Woodford-Nottingham stopping train on 9th July 1960. *Ref: MM1220*

Belgrave and Birstall

Belgrave and Birstall station was the last station before Leicester Central and followed the GCR's standard layout of being a single large island platform, this was accessed by bridge number 363 that crossed the line. It was unique as a rural station on the London Extension as it had no goods facilities due to the constraints of being built in a cutting. A signal cabin and a lamp hut were provided to the south east of the station and further afield a station master's house was constructed.

Left: A magnificent pre-nationalisation shot of Robinson O4/3 No. 3672 near Belgrave and Birstall station with a long rake of mainly wooden-bodied mineral wagons. Built by Robert Stephenson and Hawthorn Ltd, and originally taken into service by the War Department as ROD 1653 in October 1917. In December 1919 it was taken into stock by the LNWR who numbered it 2905. The loco then transferred to the LNER in June 1924 was renumbered 6354 which it carried until September 1946 when it became 3672, finally renumbered by British Railways in August 1950 to 63672 which it carried until withdrawal from 36E Repton Thrumpton (GC) in December 1963, a working life of 46 years 2 months. *Ref: NS205524A*

Above: Another Robinson 2-8-0, this time Class O4/7 No. 63761 shortly after nationalisation, travels through the deep cutting heading north, having passed through Belgrave and Birstall with a lengthy mixed goods train. *Ref: NS205538A*

Above: A late 1940s/early 1950s shot sees Thompson B1 4-6-0 No. 61111 passing through Belgrave and Birstall. *Ref: NS207027*

Below: Gresley V2 2-6-2 No. 60815 departs Belgrave and Birstall with the 6.10pm Nottingham-Leicester on 28th June 1961. The photographer is located at the site of the replacement station known as Leicester North. *Ref: MM1512*

Standard 9F 2-10-0 No. 92120 pictured passing Belgrave and Birstall signal cabin and lamp hut with a northbound parcels train on 24th July 1961. *Ref: MM1569*

GREAT CENTRAL IN THE MIDLANDS

Leicester Abbey Lane

Just a short distance from Belgrave and Birstall the line ran on a high embankment parallel to Abbey Lane (part of the A6 between Loughborough and Leicester). As mentioned previously there were no sidings at Belgrave and Birstall but there were quite substantial sidings at Abbey Lane comprising a coal depot; the sidings were also used for storing empty petrol tanks.

Above: Stanier Black Five No. 45283 pictured from allotment gardens on Abbey Lane with the semi-fast 4.25pm Marylebone-Nottingham on 17th May 1961. *Ref: MM1447*

Below: Gresley K3 2-6-0 No. 61813 on the Up 'Fish' working at Abbey Lane Sidings, Leicester on 17th May 1961, which as can be seen was also used to store carriages. *Ref: MM1448*

Raven Class B16 4-6-0 No. 61421 pictured on a Woodford Halse-York fitted freight, passes Abbey Lane Sidings, Leicester on 17th May 1961. *Ref: MM1449*

Thompson Class B1 No. 61381 at Abbey Lane Sidings, Leicester on 5th July 1961 delivering new London Passenger Transport Board (LPTB) A stock to the capital for use on the Metropolitan Line. *Ref: MM1523*

Above: Rebuilt Royal Scot No. 46156 *The South Wales Borderer* approaches Abbey Lane Sidings with the 4.38pm Marylebone-Nottingham after departure from Leicester Central on 13th July 1964. *Ref: MM2555*

Below: Standard 9F 2-10-0 No. 92074 works a Down fitted freight towards Belgrave and Birstall on 24th June 1964 and is pictured passing the post-war housing development at Mowmacre Hill which is to the north west of Leicester. *Ref: MM2512*

GREAT CENTRAL IN THE MIDLANDS
Leicester Central

Leicester Central (38C) allocated Class B1 4-6-0 No. 61163 crosses the River Soar with a freight working heading towards Leicester Central station in the mid-1950s. *Ref: AF0137*

Gresley Class A4 No. 4468 *Mallard* restored back to its former glory in LNER garter blue livery, passed through Leicester Central on 28th February 1964. The famous locomotive had been renovated at Doncaster Works and was pictured on its way to Clapham Transport Museum. The roof of the Great Central Hotel can be seen to the right of the tender. *Ref: AF1256*

The northern approaches to Leicester Central were on a higher level than the surrounding houses and streets the line being constructed on brick-retained embankments and viaducts built of Staffordshire blue brick, incorporating a series of girder bridges. Typical of the high standards to which the London Extension was built, the abutments of the girder bridges that crossed public roads were lined in white-glazed tiles to increase the level of light underneath. In total the viaduct that Leicester Central station would be constructed on was in excess of a mile and a half in length. The station opened on 15th March 1899.

As with Nottingham Victoria the viaduct's construction required a large area of land to be acquired by compulsory purchase with the GCR agreeing to rehouse at its own expense the inhabitants of around 300 houses which were to be demolished; the area principally affected by the works was the working class Blackfriars district where the slum housing was swept from the map, to be replaced by Great Central Street. Around 250 houses were constructed by the GCR in Newfoundpool to the west of Leicester.

The station was comprised within a south-west facing rectangle, bordered on the one side by Blackfriars Street and Jarvis Street, and on the other side by the new Great Central Street. The tracks ran north-east to south-west, crossing the A50 Northgate Street on a 'Bowstring' girder bridge before splaying out on either side of a large 1,245 foot H-shaped island-style platform upon which the station was built. Six running lines flanked either side of the station with bays at either end to accommodate local workings to Nottingham and Rugby. A parcels office and stabling point for locomotives were also incorporated into the site. The main station entrance was on Great Central Street where a large ornate terracotta-lined archway crowned by an ornate clock tower led through to the entrance hall and cab waiting area; the station frontage itself had a red brick and terracotta façade, to the left of which was the entrance to the parcels office. A second entrance was in Jarvis Street where a subway 20 feet below the platforms led through to the main booking hall, a light and airy space topped by a glazed roof. Stairs led up to the platforms, whilst a hydraulic lift was used to transport luggage from the booking hall.

The southern end of the new station and its viaduct required building over Jewry Wall Street and some of the houses that stood on it. In 1832 at one of these houses, number 53, a well-preserved and high-quality Roman mosaic floor was uncovered during enlargement of the cellar. The floor was preserved and the owner allowed public access to view the mosaic on request. Although number 53 was demolished, the Great Central undertook to preserve the Roman floor within the structure of the southern northbound platform that was built around it. The mosaic was encased in a brick vault topped by a glass ceiling let into the platform so it could be viewed from above. A locked doorway at street level provided access to the vault and a local shopkeeper was entrusted with the key to continue to provide access to the public upon request.

GREAT CENTRAL IN THE MIDLANDS

Above Left: Rebuilt Patriot 4-6-0 No. 45530 *Sir Frank Ree* is pictured at Leicester Central in August 1963 waiting to depart with a northbound express. *Ref: AF1284*

Below Left: A scene from Leicester Central shed (38C) sees ex-LMS Class 5MT 4-6-0 No. 44695 and ex-LNER Class B1 (also rated 5MT) 4-6-0 No. 61369 side-by-side at Leicester Great Central shed. The Black Five, in all probability, has brought 'The South Yorkshireman' from Bradford Exchange as the picture shows it was allocated at Low Moor (56F) shed, which was located just south of the west Yorkshire town. 'The South Yorkshireman' service was introduced by the Eastern Region of BR and first ran on 31st May 1948, the Up train departing from Bradford at 10.00am, arriving at Marylebone at 3.15pm. The final working of this train was on 2nd January 1960. *Ref: CW10176*

Above: Ex-WD 2-8-0 No. 90479 pictured simmering in the sidings at Leicester Central. The main station building can be seen above the rear left of the locomotive. Built at Vulcan Foundry for the War Department, entering service in June 1944 as No. 8680, the loco would acquire quite a few identities, becoming 78680 in June 1945. After being taken into LNER stock in April 1947 the number changed to 3158, it then had a 6 added at nationalisation by BR but this was never applied, eventually it became 90479 in May 1949 and carried this identity until withdrawal in September 1966. *Ref: CW10352*

Thompson B1 4-6-0 No. 61078 pauses at Leicester Central with a northbound service.
This shot affords a good view of the sidings, turntable and water tower adjacent to Great Central Road. *Ref: CW10787*

The driver of Thompson Class L1 2-6-4T No. 67758 waits patiently as parcels are unloaded from the 5.30pm Nottingham-Rugby at Leicester Central on 6th May 1960. *Ref: MM1163*

GREAT CENTRAL IN THE MIDLANDS

Robinson Class A5 4-6-2T No. 69809 pictured light engine to the north of Leicester Central station. *Ref: CW11225*

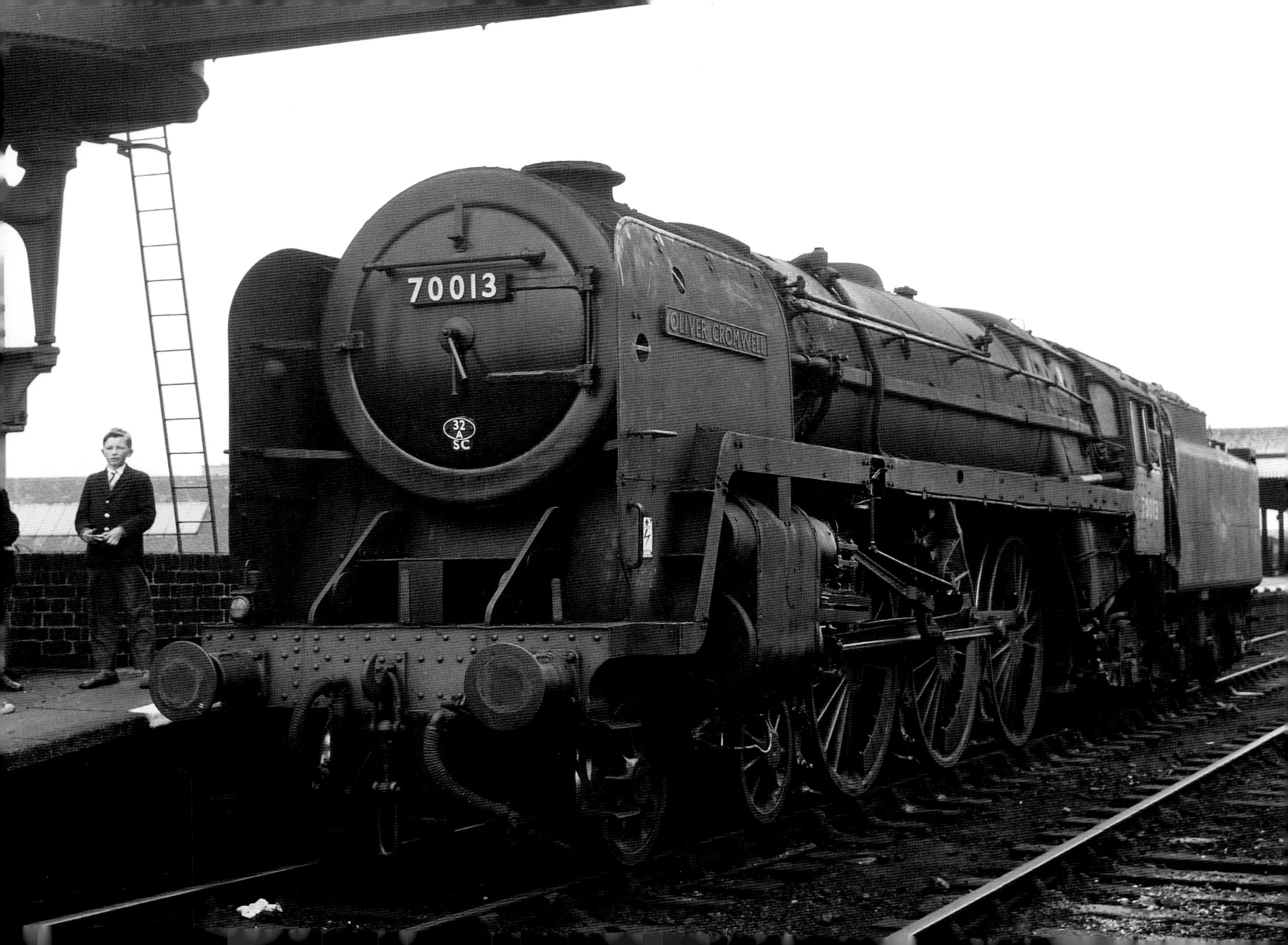

OLIVER CROMWELL
70013
32A SC

Above: Fairburn Class 4MT 2-6-4T No. 42556 arrives into the south bays at Leicester Central while Thompson Class L1 2-6-4T No. 67769 tops up its tanks prior to departure for Rugby Central on 20th August 1960. *Ref: MM1295*

Left: Britannia Class 7MT No. 70013 *Oliver Cromwell* is pictured on 23rd July 1960 pictured in one of the bays at Leicester Central. In the summer of 1968, with main line steam fast disappearing, it was the solitary Pacific locomotive remaining on the active list, as it had been the last main line engine to receive a repair at Crewe, emerging from the Works in February 1967. It returned to its Carlisle Kingmoor shed from where it was reallocated to Carnforth in January 1968 after the withdrawal of the last of its classmates. Working the occasional rail tour, it was earmarked to take the scenic Manchester Victoria to Carlisle leg of the famous 'Fifteen Guinea Special' that was to be main line steam's finale on 11th August 1968. That tour immortalised the locomotive and preservation beckoned, it was chosen to be a part of the National Collection.

The National Collection had limited storage space and took up an offer from Alan Bloom at his Bressingham steam museum in East Anglia to provide a home for the locomotive. On 12th August 1968 it ran light engine to Norwich, where many of its classmates had seen their finest hour, and then went to Diss where it completed its journey by lorry for the short distance to Bressingham, arriving there on 19th August 1968. It continued in use, albeit only on a 200 yard track, giving guards van and footplate rides. It ran until 1973 when it was placed inside the museum, where it would stay until 2004, spending twice as long there as it had been in service. *Ref: MM1249*

After leaving Bressingham Steam Museum 70013 made a star appearance at that year's Railfest in York after which she was moved to the GCR at Loughborough where a thorough overhaul took place. It returned to passenger service on 3rd May 2008. The locomotive has since gone on to operate a considerable number of steam specials on Network Rail, including the recreation of the '15 Guinea Special' on the 50th anniversary of the original train, as well as services on the preserved GCR line. At the time of writing another major overhaul is taking place. *Ref: MM1249*

Above: In stark contrast to the previous photograph we are again at the south end of Leicester Central but this time on the 20th January 1963, part way through the 'Big Freeze' that commenced on Boxing Day 1962. Blizzards and constant snowfall set the scene for the next two months, as much of England remained covered every day until early March 1963. Snowdrifts and blocks of ice were commonplace and temperatures dropped below -20°C, causing major disruption to the railways. The thaw began at the start of March when mild, south-westerly winds led to a welcome increase in the temperature. By 6th March, there were no frosts anywhere in the UK and the temperature reached 17°C. The snow quickly melted and with the thaw came flooding. Here we see rebuilt Class 7P Royal Scot No. 46163 *Civil Service Rifleman* taking water (hopefully!) while working the 10.10am Nottingham-Marylebone. *Ref: MM2047*

Top right: Rebuilt Royal Scot No. 46122 *Royal Ulster Rifleman* with a less than taxing four coach local, the 6.15pm from Nottingham to Rugby, is seen alongside Standard 9F 2-10-0 No. 92073 at Leicester Central on 3rd June 1964. *Ref: MM2484*

Below right: Standard Class 5MT No. 73053 is seen light engine after being turned and before taking over the Up 'Fish' on 6th July 1964. The loco was allocated to Woodford Halse (2F) but less than two weeks later would be transferred to Shrewsbury (6D). *Ref: MM2543*

GREAT CENTRAL IN THE MIDLANDS

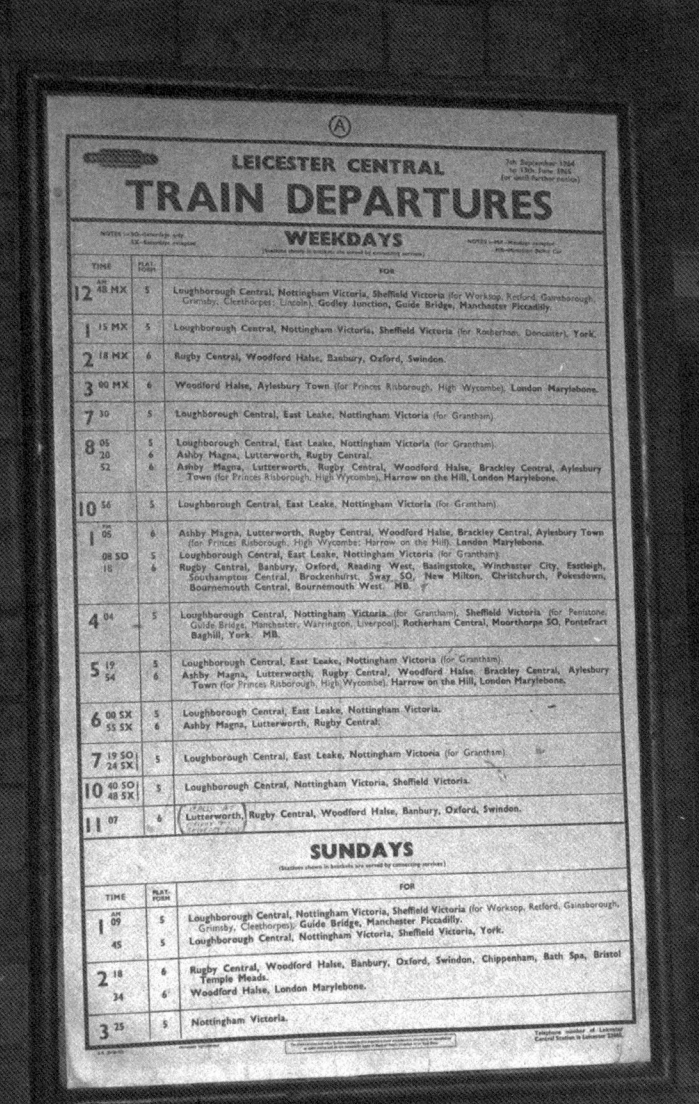

Left: A timetable of train departures from Leicester Central pictured on 22nd November 1965. The service is looking quite sparse at this time with only 8am to 9am having more than two departures, the Sunday timetable looks interesting with no departures after 3.25am! The eventual withdrawal of services would take effect from Monday, 5th May 1969. *Ref: MM3068*

Below: A couple of schoolboys chat with the footplate crew of Stanier Jubilee Class 6P No. 45676 *Codrington* as they wait for departure time at platform 5 Leicester Central with the 4.38pm Marylebone-Nottingham on 10th August 1964. *Ref: MM2632*

Right: Shortly after leaving Leicester Central, Stanier Black Five No. 45292 travels northwards on the 4.38pm Marylebone-Nottingham on 22nd June 1966. The train is crossing over the River Soar on a girder bridge which was one of many that had to be constructed on the elevated section of line through the city. The road bridge that can be seen just below the bridge deck carries the A6 into the city centre and is known as St. Margaret's Way, note the many anglers between the road and rail bridges. *Ref: MM3163*

GREAT CENTRAL IN THE MIDLANDS

GREAT CENTRAL IN THE MIDLANDS
Leicester South

Above: Thompson L1 2-6-4T No. 67760 passes Leicester South Goods with the 5.30pm Nottingham-Rugby on 30th May 1960. Leicester South Goods signal cabin can be seen above the second and third coaches. *Ref: MM1187*

Below: Two northbound services pass Leicester South Goods simultaneously, on the right an unidentified Thompson B1 4-6-0 with the 4.30pm Marylebone-Nottingham and Standard 9F 2-10-0 No. 92088 with an express freight on 14th June 1960. Leicester GC shed (38E) was located to the left of the ventilated van. *Ref: MM1197*

Leicester South Goods and the GC engine shed were hemmed in by the Midland Railway Leicester to Burton line where we see ex-LMS Class 4MT 2-6-0 No. 43018 and Class 5MT No. 45284 crossing over the Great Central on Bridge 379, note the two young lads on the right giving a wave while sitting on the wall by the nearby iron foundry. The fireman of Gresley Class K3 2-6-0 No. 61945 makes use of the time while waiting to progress to Leicester Central by dragging the coal forward in the tender. *Ref: CW10138*

Great Central in the Midlands
Whetstone

Whetstone station was opened to passengers on 15th March 1899 in the usual GCR single island platform style. Access was from beneath the railway bridge, which spanned Station Street near the centre of the village. The station was one of the earlier closures on the line closing to passengers in 1963; like nearby Ashby Magna, it served a relatively sparsely populated area and had always struggled to attract sufficient revenue.

Since closure, the railway embankment has been removed, along with all station buildings and platforms; only the Station Master's house remains today. A new housing development has been constructed on the old trackbed formation.

Above: Kicking up a dust cloud from recently laid ballast, Gresley Class K3 No. 61832 heads the 6.15pm (Sundays Only) Nottingham-Marylebone under the A426 road between Whetstone and Ashby Magna on 25th June 1961. *Ref: MM1509*

Right: Another view at the same location sees Riddles Standard 9F 2-10-0 No. 92057 working an Up ballast train on 4th November 1961. *Ref: MM1670*

GREAT CENTRAL IN THE MIDLANDS

92057

Above: Stanier Class 8F 2-8-0 No. 48700 works an Up through freight between Whetstone and Ashby Magna on 26th October 1963. *Ref: MM2319*

Above Right: A commendably clean Gresley Class V2 2-6-2 No. 60877 on an Up 'runner' pictured on Ashby Bank between Whetstone and Ashby Magna on 8th February 1964. *Ref: MM2359*

Below Right: Stanier Class 5MT 4-6-0 No. 44836 crosses the Grand Union Canal between Blaby and Whetstone with the 6.15pm Nottingham-Rugby, on 18th May 1965. The bridge (number 393) was typical of the many steel girder structures with a criss-crossed lattice parapet built on the London Extension. *Ref: MM2851*

GREAT CENTRAL IN THE MIDLANDS

GREAT CENTRAL IN THE MIDLANDS

Whetstone saw the meeting of the GC and ex-LMS Nuneaton-Leicester lines, the GC crossing over the latter. Pictured from the local Blaby to Enderby road, Stanier Class 5MT No. 44941 heads south away from Leicester with the 6.15pm Nottingham-Rugby semi-fast service on 6th June 1966. *Ref: MM3151*

Ashby Magna

Ashby Magna was situated in open countryside between the village of that name and Dunton Bassett. For the size of the station it had quite extensive sidings and facilities including a cattle dock. The construction of the M1 motorway to the east of the station occurred while the line was still open and resulted in the demolition of the Station Master's house and the loss of the goods yard.

Above: Thompson Class B1 4-6-0 No. 61106 working an Up 'Fish' train pictured between Ashby Magna and Lutterworth in early evening sunlight on 13th May 1960. The fields to the right of the train now lie under the M1 motorway. *Ref: MM1169*

Above: Standard 9F No. 92094 powers an Annesley-Woodford Halse through freight up Ashby Bank on 15th March 1961. *Ref: MM1381*

Above Right: Working bunker first Thompson Class L1 2-6-4T No. 67769 departs from Ashby Magna with the 5.30pm Nottingham-Rugby on 2nd May 1961. *Ref: MM1424*

Below Right: Gresley Class V2 2-6-2 No. 60864 powers a Down freight through Ashby Magna and is about to pass Annesley-based Stanier 5MT 4-6-0 No. 45217 with an FA Cup special working (M672) from Leicester Central to Marylebone on 6th May 1961. The journey home for Leicester City supporters would be a sombre one as they lost 2-0 to Tottenham Hotspur who completed the double after also winning the First Division title. *Ref: MM1433*

GREAT CENTRAL IN THE MIDLANDS

Above: Gresley Class V2 2-6-2 No. 60898 passes through Ashby Magna with the 10.00am Bradford-Poole on 24th June 1961. *Ref: MM1507*

Above Right: BR Standard 9F 2-10-0 No. 92073 of Annesley shed (16D) is pictured south of Ashby Magna having just passed through Dunton Bassett Tunnel on 11th July 1962. The tunnel was only 92 yards in length and was more commonly referred to as Ashby Tunnel. *Ref: MM2044*

Below Right: The effects of the freezing cold winter of 1962/63 are all around as rebuilt Royal Scot No. 46101 *Royal Scots Grey* hurries along with an Up parcels train having passed through Ashby Magna station on 19th January 1963. *Ref: MM2044*

GREAT CENTRAL IN THE MIDLANDS

Above: Stanier Jubilee Class 6P No. 45622 *Nyasaland* looks to be moving along nicely with the 7.15am Leicester Central-Marylebone FA Cup Final Special (1X50) south of Ashby Magna on 25th May 1963. The FA Cup Final would usually be played at the beginning of May but due to the harsh winter of 1962/63 many games were postponed leading to an extended season. Leicester City would once again fail at the final hurdle, this time losing 3-1 to Manchester United. *Ref: MM2142*

Top Right: Riddles Standard Class 4MT 2-6-0 No. 76086 departs Ashby Magna with the 6.15pm Nottingham-Rugby local service on 9th August 1963. The loco doesn't show up as being allocated to any former GC shed and at the time the photograph was taken is listed as being at Trafford Park (9E). We also now see that the construction of the M1 motorway is encroaching upon the railway and around the goods yard and the station masters house in the background, which would eventually be demolished. *Ref: MM2257*

Below Right: Standard 9F 2-10-0 No. 92092, not looking in the best of health, leaking steam from a few places, takes its Up through freight over Ashby Bank on 7th December 1963. *Ref: MM2332*

GREAT CENTRAL IN THE MIDLANDS

Above: A rare sight on the Great Central Main Line sees Stanier Class 8P 'Coronation' Pacific No. 46251 *City of Nottingham* working the first leg of the RCTS 'The East Midlander No. 7' rail tour near Ashby Magna on 9th May 1964. The tour covered almost 360 miles in the day with 46251 departing Nottingham Victoria at 7.35am, taking the train to Didcot via Culworth Junction, Banbury and Oxford. Here the train was taken forward to Eastleigh station by Bulleid West Country Class No. 34038 *Lynton*. A tour of Eastleigh Works was next on the itinerary with USA Class 3F 0-6-0T No. 30071 taking over for the short journey from the station; passengers were able to spend the best part of a couple of hours before departure from the Works yard behind 34038. The return took the train to Swindon Works Junction via Salisbury. No. 46251 would then return the tour to Nottingham Victoria where it arrived at 22.46. A sign of the popularity of the tour is that it was originally booked for eight coaches but actually ran with 12. *Ref: MM2435*

Above Right: English Electric Type 3 D6740 at the head of a Rugby League Challenge Cup Final Special (1X56) near Ashby Magna on 9th May 1964. The train was taking Hull Kingston Rovers fans to Wembley where they would play Widnes who would return north victorious after beating Rovers 13-5. *Ref: MM2440*

Below Right: Standard 5MT 4-6-0 No. 73156 calls at a run down Ashby Magna station with the 6.15pm Nottingham-Rugby on 5th June 1964. The construction of the M1 progresses with a bridge being built to take the road between Ashby Magna and Dunton Bassett over the motorway. Of note is the elevated siding to the right of the locomotive, presumably installed to take spoil away from the site. *Ref: MM2487*

GREAT CENTRAL IN THE MIDLANDS

Two years later and with the motorway bridge now complete Stanier 5MT No. 45288 with three smart ex-LMS coaches in tow heads away from Ashby Magna station with the 6.15pm Nottingham-Rugby on 9th May 1966. *Ref: MM3112A*

Just a few days away from complete closure of the remaining section of the GCR main line a Class 108 two-car DMU passes through the closed station at Ashby Magna on 1st May 1969 with the 6.55pm Rugby Central to Nottingham Arkwright Street service. The tidy looking M1 in the background contrasts starkly with the run down railway. *Ref: MM3583*

Lutterworth

Lutterworth station was built to the typical GCR island platform layout, access was from beneath a bridge over a bridleway at the end of Station Road from where steps led up to the booking office at the south end of the platform. The goods yard was sited on the Down side of the line alongside the station and had quite a large signal cabin. The additional capacity in the lever frame indicated that the goods yard was planned with expansion in mind, although this did not happen.

The site of the goods yard is now occupied by Boundary Road and the station site is occupied by a housing estate. An overgrown remnant of the platform survives on the bridge at the south end of the station. The bricked up entrance also survived beneath the bridge. The Station Master's house survived and is now in private occupation.

Above: Super power for an express freight in the shape of double-headed Standard 9F 2-10-0s Nos. 92073 and 92091 pictured north of Lutterworth on 17th August 1960. *Ref: MM1289*

GREAT CENTRAL IN THE MIDLANDS

Shawell

The line ran just to the west of Shawell in the Harborough district of Leicestershire. Situated mainly in a deep cutting where Shawell signal cabin was also sited. Although there was never a station at Shawell, one was proposed a little way to the south-west where the line crossed over the A5. The cutting is now partly filled in.

Above: Thompson Class B1 4-6-0 No. 61192 works the 9.30am (SuO) Sheffield-Swindon near Shawell signal cabin on the bright winter's morning of 14th January 1962. *Ref: MM1712*

Above left: Thompson Class B1 4-6-0 No. 61269 with the 9.50am Marylebone-Nottingham comprised of vintage coaching stock has just departed Lutterworth station on 22nd July 1962. The north end of the island platform can be seen in the background as can an unidentified Riddles Austerity 8F 2-8-0. *Ref: MM1879*

Below left: Colwick-based Stanier Black Five No. 45454 departs Lutterworth with the 8.15am Nottingham-Marylebone on 12th March 1966. *Ref: MM3088*

GREAT CENTRAL IN THE MIDLANDS
Newton

We now enter Warwickshire where the line passed through the small village of Newton, as with Shawell no station existed here, but it was notable for a high three-arch bridge that carried the northern road out of the village and over the GCR cutting.

Above: Gresley Class K3 2-6-0 No. 61977 is seen on 12th February 1961 at the head of an engineering train attending the site of a derailment at Newton in the early hours of the previous morning. The track gang can be seen at the rear of the train repairing the extensive damage to the permanent way. *Ref: MM1355*

The official Ministry of Transport report introduces the incident thus: *The following is the result of my inquiry into the derailment and subsequent collision that took place shortly before 3am on 11th February on the Down and Up lines respectively between Rugby (Central) and Lutterworth stations on the Great Central line in the London Midland Region, British Railways.*

The 1.50am express freight train from Woodford to Mottram, travelling on the Down line became divided after the derailment of a wagon, and the rear portion came to rest with the leading wagon blocking the Up line. A few minutes later at about 2.48am the 10.23pm express passenger train from York to Swindon on the Up line collided at speed with the derailed wagon. The engine and tender turned on their sides and the tender turned end for end, becoming separated from the engine. The coupling came away between the tender and the leading vehicle which overran the tender diagonally to the left before it came to rest in the field adjoining the railway. Many of the other vehicles were derailed but there was no telescoping and no destruction of bodywork except to the first vehicle, which fortunately was not a passenger coach.

Two of the 18 passengers in the train were slightly injured. I regret to report, however, that Driver A. L. L. Jones was trapped on the engine and received fatal injuries; his fireman and two other members of the train crew, and the guard of the freight train, suffered from shock.

Above right: Thompson B1 No. 61077 runs tender first with an empty ballast train passing classmate No. 61373 on the 9.30am Sheffield-Swindon at Newton on 28th January 1962. *Ref: MM1722*

Below right: Riddles Standard Class 7MT Britannia No. 70040 *Clive of India* passes underneath the fine three-arched bridge at Newton on 7th July 1962 at the head of the 5.20pm (SO) Woodford-Nottingham Victoria. *Ref: MM1859*

GREAT CENTRAL IN THE MIDLANDS

GREAT CENTRAL IN THE MIDLANDS

Rugby Central

Standard Class 9F No. 92013 makes for a magnificent sight while crossing the girder bridge that passed over the West Coast Main Line at Rugby. The date was unrecorded. *Ref: W11347*

Rugby Central station was situated on Hillmorton Road, about half a mile east of the town centre. It had services between London Marylebone and Manchester Piccadilly via Leicester Central, Nottingham Victoria and Sheffield Victoria, as well as various inter-regional services to places such as Southampton and Hull.

Rugby Central was roughly midway along the Great Central Main Line (GCML) and was a stopping point for express services, as well as a changeover point for local services. Until the early 1960s, the station was served by about six London to Manchester expresses daily, and was the terminus for local services from Aylesbury or Woodford Halse to the south, and Leicester Central or Nottingham Victoria from the north.

Most of the GCML was closed on 5th September 1966, following the recommendations of the Reshaping of British Railways report. On that date, the line south of Rugby Central and north of Nottingham Victoria was closed. Until 3rd May 1969, the section between Rugby Central and Nottingham (initially Nottingham Victoria, later cut back to Nottingham Arkwright Street) remained open as a self-contained branch, providing DMU-operated local passenger services.

The station formally closed on 5th May 1969 along with what remained of the line. The station buildings were demolished after closure, but the platform still exists and is open to the public. The station site, and 4½ miles of the former Great Central Railway trackbed through Rugby, are now owned by Rugby Borough Council, who bought them in 1970 for £5,500. The trackbed runs mostly through cuttings, and it is now used as a footpath, cycleway and nature reserve called the Great Central Walk.

The former goods yard was west of the station and was used as a timber yard until the mid-1990s, when houses were built on it.

Above: Thompson Class B1 4-6-0 No. 61381 shortly after arrival at Rugby Central from Leicester Central on 2nd July 1960.
Ref: H1604

Ex-GWR Collett Hall Class 4-6-0 No. 6952 *Kimberley Hall* passing Rugby Station signal cabin with an inter-regional express some time in 1957. *Ref: MM185*

It wasn't unusual to see Standard 9Fs pressed into service on passenger trains and an example is shown here with No. 92031 heading the 2.38pm Marylebone-Nottingham express northwards between Rugby and Lutterworth on 15th August 1964. *Ref: MM2666*

GREAT CENTRAL IN THE MIDLANDS

Plenty of spotters in attendance to observe an unidentified Standard 9F crossing the West Coast Main Line on the imposing Rugby Viaduct and about to enter the girder section. *Photo: Ian Mackenzie*

Stanier Black Five No. 44858 gets a top up of water at Rugby Central before continuing its journey in September 1966. *Ref: WS8771*

Braunston & Willoughby

Braunston and Willoughby station served the small village of Willoughby which it was located next to, and the larger but more distant village of Braunston. The entrance was located on the A45 Coventry to Daventry road that passed beneath the line.

Known originally as Willoughby for Daventry which was some five miles to the south east in Northamptonshire and it already had a station of its own on another line (the LNWR branch line from Weedon to Leamington Spa). Braunston, also in Northamptonshire, lay between the two, some two miles away and also served by the same LNWR branch that ran through Daventry, but it was Braunston that was found to be providing the new Great Central station with the bulk of its usage. This was reflected in a renaming on 1st January 1904 to Braunston and Willoughby for Daventry, the name Daventry was eventually dropped in 1938.

The station was an early closure, this happening for both passengers and goods traffic on 1st April 1957. The station buildings had already been removed in 1961-2 although the platform remained for a while longer. Today there is little left to see at the site. The twin bridges over the A45 have been removed and the abutment walls substantially lowered. The Station Master's house remains, however. A short distance to the south of the station was the 13-arch Willoughby Viaduct crossing the River Leam, but as with the railway it has been removed.

Above: Stanier 5MT 4-6-0 No. 45223 hurries the 12.25pm Marylebone-Nottingham through the disused Braunston & Willoughby station on 8th July 1961. Surprisingly for a station that had been closed for four years the substantial running-in board still remains in situ. *Ref: MM1531*

Above Left: Raven Class B16/2 4-6-0 No. 61475 working a Down freight of wooden and steel bodied mineral wagons past the quite sizeable Braunston & Willoughby signal cabin on 26th August 1961. Although not unusual it was not too common to see B16s working this far south on the GC. *Ref: MM1627*

Below Right: With the station buildings now demolished Standard 9F No. 92087 rumbles through the island platform with an Annesley-Woodford Halse 'runner' on 14th April 1962. *Ref: MM1772*

GREAT CENTRAL IN THE MIDLANDS

97

Staverton Road

Staverton Road signal cabin was located between Braunston & Willoughby and Charwelton stations and sat on an embankment adjacent to the A425 in Northamptonshire. The cabin was located here to break up what would have been a 7 mile block section, it had a 20-lever frame of which 13 were in use. A refuge siding on the Up side was added by the LNER in 1923.

Above: Standard 9F 2-10-0 No. 92012 storms past Staverton Road signal cabin on 27th May 1961. *Ref: MM1460*

Above Right: Thompson Class B1 4-6-0 No. 61028 *Umseke* heads the 10.10am Newcastle-Swansea inter-regional service and is pictured near Staverton Road on 8th July 1961. *Ref: MM1538*

Below Right: Shortly after passing the signal cabin, the line crossed the Staverton Viaduct. Built in 1897 by T. Oliver & Son (who were the contractors for the Rugby-Woodford section), it measured 119 yards and comprised nine arches. A four-car DMU has just crossed over the structure with the 12.30pm Nottingham-Marylebone on 18th July 1964. *Ref: MM2569*

GREAT CENTRAL IN THE MIDLANDS

Gresley Class V2 2-6-2 No. 60941 pictured near Staverton Road on 9th May 1964. *Ref: MM2454*

GREAT CENTRAL IN THE MIDLANDS

Catesby Viaduct and Tunnel

The 12-arch Catesby Viaduct crosses the infant River Leam and located approximately 1,100 yards south is the northern portal of Catesby Tunnel, which at 2,997 yards long is the longest on the London Extension. In terms of both length and gauge, Catesby Tunnel is unusually large, at 27 feet wide and 25 feet 6 inches high, it is straight throughout and runs on a rising gradient of 1:176 towards the south. Constructed of Staffordshire Blue brick (around 30 million of them), the tunnel was completed in 1897 taking just over two years from the sinking of the first shaft, and was closed in 1966. After lying abandoned and flooded for over 50 years, proposals were granted in 2017 for the conversion of the wide, straight tunnel into an aerodynamic test facility for road and race cars.

Above: With DMUs now replacing withdrawn steam locomotives a four-car example is pictured on Catesby Viaduct with the 8.38am Marylebone-Nottingham on 17th April 1965. The tunnel entrance can be seen in the background. *Ref: MM2804*

GREAT CENTRAL IN THE MIDLANDS

Charwelton

Standard Class 5MT No. 73156 on the 5.15pm Nottingham-Marylebone passes through Charwelton on 27th May 1961. *Ref: MM1467*

English Electric Type 3 D6800 at Charwelton with an inter-regional express from Bournemouth-York on 25th May 1963. Released into traffic in December 1962 the loco is still working 60 years later for Network Rail as 97 301. *Ref: MM2148*

As can be seen from these photos Charwelton had quite an extensive goods yard. This was to serve the ironstone quarry at Hellidon where in 1917 a 1½ mile line was constructed to link with the GC. At its peak the goods yard was busy with up to 200 wagons stabled in its sidings at any one time. Both quarry and line closed on 18th November 1961, the branch being dismantled in June 1963 and the sidings at Charwelton following in 1964. The station closed to passengers and goods on 4th March 1963. Here we see an unidentified Standard 9F with an Up freight passing the signal cabin and station on 27th May 1961. *Ref: MM1464*

Woodford Halse

Above: Riddles Austerity 2-8-0 No. 90346 enters Woodford Halse station at 1.43pm on 2nd July 1960. *Ref: H1580*

The station at Woodford opened under the name Woodford and Hinton and served the adjacent villages of Woodford Halse to the east and Hinton to the west, both in Northamptonshire. The station was renamed Woodford Halse on 1st November 1948.

The village of Woodford Halse became notable for the role it played as an important railway centre. Originally it had seemed the railway would by-pass Woodford, as there were stations at Byfield (two miles west), and Moreton Pinkney (three miles south east) which were both on the East and West Junction Railway (becoming part of the Stratford-upon-Avon and Midland Junction Railway – (S&MJR)) and no other lines seemed likely to be built in such a thinly populated area. However, in the late 1890s the village found itself on the Great Central Railway's London Extension.

The station was a variation on the standard island platform design typical of the London Extension, here the less common 'embankment' type reached from a roadway that passed beneath the line. It differed from the usual design in that since it served what was effectively a four-way junction, it was provided with a more extensive range of platform buildings and facilities.

The station was situated near Woodford Halse North Curve Junction linking the GCR with the S&MJR route between Stratford-upon-Avon and Towcester, a separate platform was provided on the west side for these trains, a timber structure later replaced by a concrete slab construction (as in the picture above) although staff still referred to it as the 'wooden platform'. Further south however was the more important Culworth Junction, a divergence point for a stretch of line 8¼ miles in length linking the GCR with the Great Western Railway at Banbury enabling extensive and varied cross-country workings to take place.

North of the station was the locomotive depot, wagon and sheet repair shops, plus extensive marshalling yards. These facilities were originally intended to be located at Brackley. Local opposition forced the GCR to change its plans and the site moved to Woodford Halse. Located on top of a vast embankment the site covered around 35 acres formed mainly from spoil taken from Catesby Tunnel. 136 terraced dwellings to house the railway workers were built on the east side of the embankment, together with a street of shops. This gave a small rural village an industrial look and increased the population to around 2,000. The village had a railway workers' social club and cinema amongst its facilities.

Woodford Halse depot and yards were a hive of activity, but not busy enough to ensure survival. On 5th April 1965 the marshalling yards closed, and on 5th September 1966, most of the GCR was closed completely, including all remaining lines converging on Woodford Halse.

A view looking north to Woodford Halse station with the south junction trailing off to the left. An unidentified Standard 5MT approaches the station on 3rd March 1957. *Ref: MM202*

GREAT CENTRAL IN THE MIDLANDS

Above Left: Stanier Black Five No. 44830 on the 3.8pm Woodford-Marylebone passing the S&MJR Woodford South curve on 1st March 1959. *Ref: MM758*

Below Left: Stanier Class 5MT No. 45091 has departed Woodford Halse with the RCTS 'Grafton Rail Tour' on 9th August 1959. The tour originated at Kings Cross with North British Type 2 diesel D6101 taking the special up to Hitchin, steam power then took over in the shape of veteran Johnson Class 3F 0-6-0 No. 46474 for the journey to Bedford. From there the first of three different Stanier Black Fives was used, No. 45139, which took the train to Blisworth. The loco in our photograph then travelled via Calvert Junction and Verney Junction to Banbury Merton Street. Passengers then had to walk to the nearby Banbury General where ex-GWR 57XX pannier tank No. 3646 took the train to Leamington Spa South Junction. The final Black Five of the day, No. 44883 went forward to Luton Bute Street where it handed over to the original diesel locomotive to return the tour to Kings Cross, quite a feat of organisation! *Ref: MM983*

Above: An unidentified 2-6-4 tank witnesses the departure from Woodford Halse of Stanier 5MT 4-6-0 No. 45417 on the 2.38pm Marylebone-Nottingham service on 25th May 1963. *Ref: MM2146*

GREAT CENTRAL IN THE MIDLANDS

Culworth

Standard 5MT 4-6-0 No. 73157 on the 3.5pm Woodford-Marylebone calls at Culworth on 3rd March 1957. Bridge 508 in the background looks grand with each of the three spans measuring 55 feet 6 inches; it was though a humble occupation bridge. The whole scene looks clean and tidy and built to the highest standards. *Ref: MM203*

Above: Stanier Black Five No. 44872 on the 8.15am Nottingham-Marylebone south of Culworth on 3rd September 1966.
Ref: MM3279

The station was located midway between the stations at Woodford & Hinton and Helmdon near the village of Moreton Pinkney in Northamptonshire, this name could not be used because there was already a station with this name by the village served by the Stratford-upon-Avon and Midland Junction Railway (S&MJR) which was built 26 years earlier, in 1873. Instead, the name of the next nearest significant village was chosen, Culworth, which was ¾ mile away.

A year after opening, a branch was built between nearby Woodford Halse and Banbury and in 1913, a small station was added on the western edge of Culworth, this was named Eydon Road Halt.

The traffic at Culworth was of the rural kind and relatively light with only local services calling at the station, so if travelling to the larger towns on the line would necessitate a change at Woodford Halse or Brackley. Goods traffic was handled by pick-up trains between Woodford Halse and Quainton Road or Marylebone. As both passenger and goods traffic declined Culworth became one of the first stations on the GC Extension to be closed, this happened on 24th September 1958 for passenger traffic, and finally to all traffic on 6th June 1962.

Thorpe Mandeville

The Banbury branch left the GC main line at Culworth Junction linking with the Great Western Line at Banbury Junction, a distance of just over 8 miles. This enabled inter-regional services access to the north and resulted in a large variety of locomotives to pass over the GC route. There were two small halts on the line, Eydon Road (for Thorpe Mandeville) and Chalcombe Road (for Chacombe) – note the difference in the spelling between the halt and the village – they were both served by a local link service from Woodford which was known as the 'Banbury Motor', both halts would close in 1956.

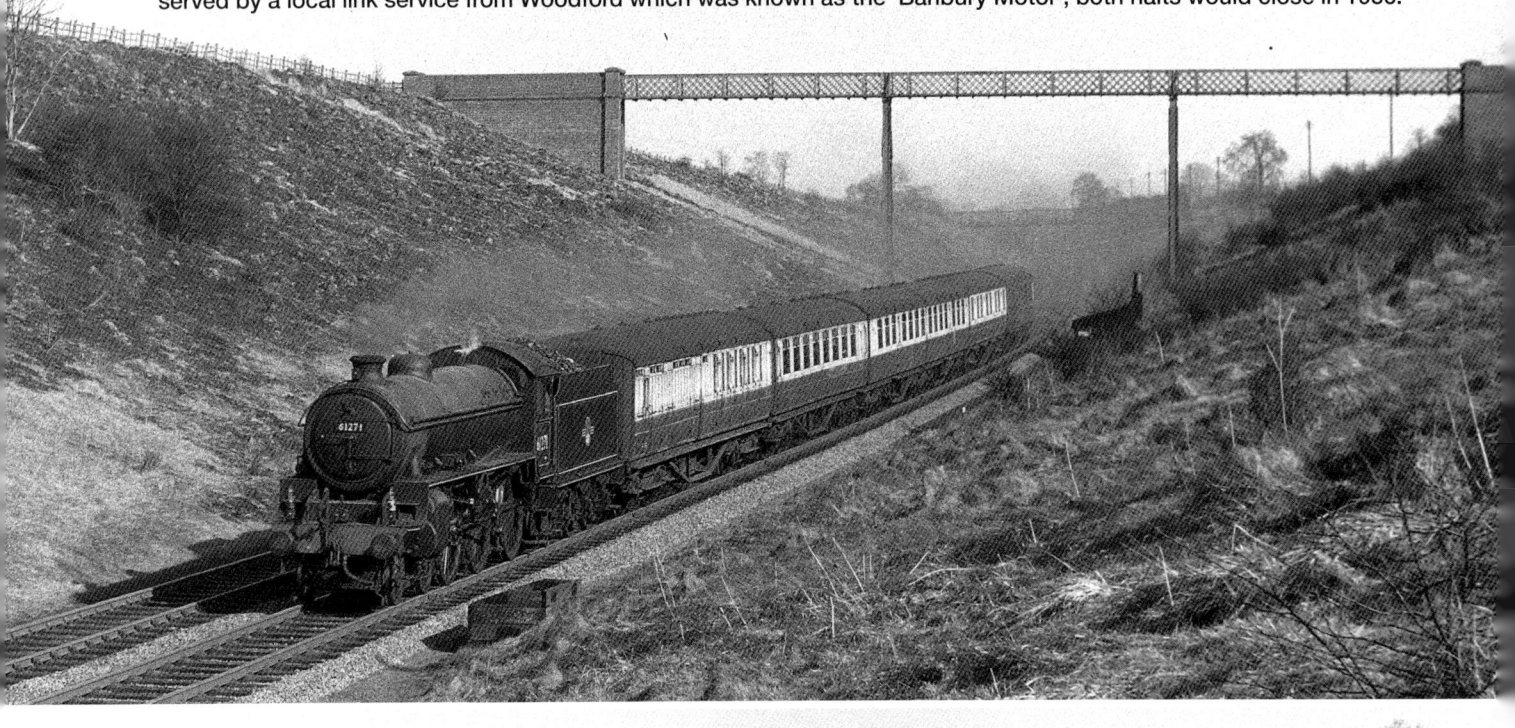

Above: Thompson Class B1 No. 61271 passes through the deep cutting west of Thorpe Mandeville with the 1.5pm Woodford-Swindon on 1st March 1959. *Ref: MM757*

Below: Gresley Class V2 2-6-2 No. 60863 pictured near Thorpe Mandeville working the 8.10am Swansea-York on 4th April 1959. *Ref: MM904*

Helmdon

A typical GC island construction, the station lay south of a nine-arch viaduct that crossed over the River Tove. Helmdon was the nearest station for Sulgrave Manor, which had been the home of George Washington's ancestors in the 16th and 17th centuries. In the 1920s the house was restored and opened as a museum, so the station was renamed Helmdon for Sulgrave in 1928 by the LNER. The station closed to passengers on 4th March 1963 and to goods on 2nd November 1964. The platforms remain but are now largely hidden in the undergrowth. The viaduct remains.

Above: Looking a credit to the station staff, a pristine Helmdon station greets Thompson B1 No. 61138 as it passes through with the 1.5pm Woodford-Marylebone, on 11th June 1962. *Ref: MM1835*

Brackley Central

Gresley Class V2 2-6-2 No. 60879 powers 'The South Yorkshireman' north through Brackley Central on 4th July 1959. As with Helmdon the station and surrounds are commendably clean and tidy, which is more than can be said for the exhaust being emitted from the V2's chimney. Ref: MM917

We finish our journey south through the Midlands at Brackley Central, close to the Northamptonshire border with Oxfordshire and Buckinghamshire. It was the second station to serve the town, the other being the LNWR's station, simply known as Brackley; this station closed in 1964, Central closed to all traffic on 5th September 1966.

The station was the more common 'cutting' type reached from the A43 which crossed over the line. The cutting itself was substantial and typical of the line, involving the excavation of around 336,000 cubic yards of material. The original intention was for the station entrance to be located on the bridge itself, as it was at most other stations on the line, but local concerns about traffic congestion forced a change in the layout, the entrance being relocated at the top of the cutting on the west side in a lay-by off the road, with a footbridge leading across to the island platform. Although the town of Brackley had a population of barely 2,500 at the time, it was considered a sufficiently large and important settlement for the station to be provided with a more extensive range of platform buildings and facilities, as at Woodford Halse and Rugby Central. It is thought that the GCR intended to build a link to Northampton and it would appear that provision was made for an additional platform.

South of the station the line crossed the Great Ouse by means of a 22-arch viaduct measuring 320 feet long necessitating large embankments on either side. There were problems at the southern end during construction due to moving beds of clay and two arches were removed and replaced by girder spans with a third arch being converted to a buttress to act as support to the last two spans and help stabilise the structure. Unfortunately the viaduct was demolished in 1978.